DIESELS over DONNER

MOUNTAIN SOUL OF THE SOUTHERN PACIFIC

"Fighting Back"

This book is dedicated to Philip Anschutz and the SP and Rio Grande people bringing a new spirit to the West.

All Uncredited Photos
Richard Steinheimer

Design and Production
James Winter Reese

Typography
Roc-Pacific, Los Angeles, CA
Express Type Co., Simi Valley, CA

Printing and Binding
Walsworth Publishing, Marceline, MO

DIESELS *over* DONNER

MOUNTAIN SOUL OF THE SOUTHERN PACIFIC

Sacramento Locomotive Works 1985

BY RICHARD STEINHEIMER and DICK DORN
with Shirley Burman and Donald Sims

INTERURBAN PRESS
Glendale, California 91225

"Daylight" SD40R No. 7342 shoves through Emigrant Gap

Diesels over Donner
Mountain Soul of the Southern Pacific

Library of Congress Cataloging-in-Publication Data

Steinheimer, Richard.
Diesels over Donner : mountain soul of the Southern Pacific / by Richard Steinheimer and Dick Dorn : with Shirley Burman and Donald Sims.
p. cm.
Includes bibliographical references.
ISBN 0-87046-090-0
1. Southern Pacific Railroad Company. 2. Railroads—California—Donner Pass. 3. Diesel locomotives—United States. I. Dorn, Dick. II. Title.
TF25.S68S74 1989
385'.09794—dc20 89-37339
CIP

Table of Contents

Introduction

Today's version of Theodore Judah's 1860 plan for a railroad over Donner Pass would be a monomainiac space freak convincing four Des Moines shopkeepers to construct a hotel on the Moon."

This book was envisioned several years ago with Dick Dorn as a diesel age "pictorial supplement" to the book, *Donner Pass,* that well researched and illustrated history by John Signor.

Diesels Over Donner is a vivid and personal pictorial of the fascinating parade of Southern Pacific trains across the West's first great mountain railroad. It's a visual history of first and second generation diesel locomotive operations before they are supplanted by a third diesel generation. In these pictures we see the mountains and the powerful forces of nature that helped shape the soul of today's Southern Pacific.

More than 20 SP diesel classes are pictured, several specifically designed for the tunnels and snowsheds of this grade. The photographs of trains and scenery will interest railfans, modelers, railroad employees and people who simply want to learn about contemporary railroading on this historic mountain pass.

The idea for a color book originated with seeing the excellent color photographs of Dick Dorn and other photographers made in the heavy Donner storms in the winter of 1981-82. That was a time of hard work for me and my associate, Curtis Fukuda, in our new commercial photography studio in Mountain View. Though I had done pictures on Donner since 1951, I'd never been able to get into the mountains at the right time for such excellent storm pictures. The result was a personal vow to get into the mountains the next winter— business or no business.

What luck! The winter of 1982-83 was the second heaviest snow season in Sierra history. Meteorologist Mike Pechner introduced me to Charley Drinnon, maintenance of way chief in Truckee, who showed me how SP railroaders could keep the pass safe and open despite an incredible snow bonanza.

From that came contacts with more SP railroaders, including assistant division engineer, Jim "The Bear" Mahon, and his experienced boss, Bill Lynch. Trainmaster Gerald Nester and other operating officers and trainmen were willing to explain things. Shirley Burman and I photographed bridge and building foreman Dick Carter as he constructed a new snowshed. Shorty Teal, Earl Stice, Harry Hanns and other operators talked about life in the little community of Norden where the average annual snowfall totals about 37 feet.

Every winter the snow crews face their equivalent of the Super Bowl— a yearly fight against the same tough opponent that skunked the Donner Party and nearly defeated the builders of the transcontinental railroad. Like pro football players, they know that winning can be threatened by even the most insignificant seeming decision or careless act on one person's part— unless everybody acts professionally— as if the entire job depends on them.

The result is a very safe railroad with a better record for staying open than adjacent Interstate 80.

Influenced by this safe and sane approach to a potentially dangerous job, we created the color photography show entitled, "Winter's Professionals", which premiered at the California State Railroad Museum on March 1, 1986. Dick Dorn, Shirley Burman and Dale Sanders contributed incredibly fine photographs for the exhibition, some of which appear in this book. The pictures were taken with the professional attention to safety demanded of the railroaders, but soft, deep snow often made photography as hard as shoving a camera in the face of the Devil.

Diesels Over Donner provides a broad geographic coverage of contemporary operations, as well as many of the finest photographs taken on The Hill in the 1980's decade. In the style of *Backwoods Railroads of the West*, the book attempts to tie together unique visual images with facts and personal observations, to tell the story of railroading in this soul country of the Southern Pacific with a human scale.

The photographs from the recent "terrible 1980's" help us better understand historic events. They show us what the deep snows looked like that stranded the Donner Party in 1846 and tortured the Chinese and Irish railroad construction workers in the mid-1860's as they constructed the Central Pacific. This book is a link between the men who built this railroad that was expected to be "lost in the clouds of the Sierra Nevada" and today's inheritors of the pass.

As a young man back in the 1950's, I conceived the idea of doing a great photo story of railroad life at Sherman, Wyoming, that raw little community at the top of Sherman Hill on the Union

Pacific mainline west of Cheyenne. I could visualize pictures of the great steam locomotives flashing past, and the faces of the train order operators and other railroaders out in the incessant winds and storms of the pass. Though the idea was good, it was a project way beyond my financial and organizational abilities. Soon, steam locomotives and train order operators disappeared from Sherman and I forgot that youthful dream.

Then one day, looking at the large color prints Shirley Burman and Arlene Palmer were framing for "Winter's Professionals", it suddenly dawned on me that everything I had once wanted to do on Sherman Hill had been accomplished in three years of work with the railroaders of Donner Pass.

It was a wonderful realization. For that we all have the Winter's Professionals of Donner Pass to thank.

Richard Steinheimer
Sacramento, CA

To Learn More About Donner Pass

Beebe, Lucius, *The Central Pacific & The Southern Pacific Railroads*. Berkeley, Howell-North Books, 1963.

Best, Gerald M., *Iron Horses To Promontory*. San Marino, Golden West Books, 1969.

Best, Gerald M., *Snowplow*. Howell-North Books, Berkeley, 1966.

Crofutt, George A., *New Overland Tours and Pacific Coast Guide*. Omaha, The Overland Publishing Co., 1880.

Fisher, Kay, *A Baggage Car With Lace Curtains*. B&K Fisher, Colfax, 1979.

Garmany, John Bonds, *Southern Pacific Dieselization*. Edmonds, PFM Publications, 1985.

Heath, Erle, *Seventy-Five Years of Progress*. San Francisco, Southern Pacific, Dec. 1945.

Hinckley, Helen, *Rails From The West*. San Marino, Golden West Books, 1969.

Hoppe, Robert B.,"Waiting for the Sun", *Southern Pacific Bulletin* pp 14. San Francisco, Southern Pacific Transportation Co. June 1983.

Hoppe, Robert B.,"Snow Sheds in the Sierra", *Southern Pacific Bulletin* pp 18. San Francisco, Southern Pacific Transportation Co. Dec. 1984.

Hunt, Thomas H., *Ghost Trails To California*. Palo Alto, American West Publishing Co., 1974.

Jewell, Donald V., *Southern Pacific Motive Power Annual 1977-80*. Burlingame, Chatham Publishing Co., 1980.

Kraus, George, *High Road To Promontory*. Palo Alto, American West Publishing Co., 1969.

Lewis, Oscar, *The Big Four*. New York, Alfred A. Knopf, Inc., 1938.

Mills, Ev, *Overland Route*. Sacramento, History West, 1981.

Shine, Joseph W., *SPSF Motive Power-1986*. La Mirada, Four-Ways-West Publications, 1986.

Signor, John R., *Donner Pass*. San Marino, Golden West Books, 1985.

Stewart, George R., *Ordeal By Hunger*. Houghton Mifflin Co., New York, 1936.

Stewart, George R., *Storm*. New York, Random House, 1941.

Stewart, George R., *To California By Covered Wagon*. New York, Random House, 1954.

Wilson, Neill C. and Taylor, Frank J., *Southern Pacific*. New York, McGraw-Hill Book Company, Inc., 1952.

Roseville- February 1986

SP
9211 9211
SP
8338 8338
SP

Sacramento Heartland

"Theodore Judah was considered one of the great crackpots of Sacramento, always insisting he knew how to lay a transcontinental railroad across the Sierra Nevada."

The Central Pacific should have been born as a wealthy industrial creation of the city of San Francisco, arching steel rails eastward from the Golden Gate to a glorious union with the Industrial East. But the ridicule of Theodore D. Judah's dream by the City's financiers closed that door, casting the Central Pacific, and today's Southern Pacific, into the role of railroads which would always have to fight for survival.

Judah died before he could get East Coast financing so his plan was carried on by four Sacramento shopkeepers who gained wealth from the Gold Rush. Backed by Federal subsidies lobbied through Congress by Judah, the merchants' dreams would often be more of achieving position and monopoly than benefiting the West with a great railroad.

Those controlling partners of the Central Pacific were Collis P. Huntington, Mark Hopkins, Leland Stanford, and Charles Crocker. They would become known as The Big Four, men who for three decades would control most of the politics and commerce of California.

Carried forward by these unlikely foster parents, the first spike of the transcontinental railroad was driven in 1863 at the foot of K Street on the Sacramento waterfront, about 15,000 miles away by sea from the nearest major suppliers of steel and railroad equipment. That same year, work began nearby at what would become the great Sacramento Shops.

Even the sanity of The Big Four was seriously doubted as they followed "Crazy Judah's" dream of a railroad across the Sierra Nevada— until the day they moved the entire range 20 miles closer to Sacramento.

That happened when the rails reached Arcade Creek, only seven miles across the flat valley plain from Sacramento. Influenced by some hidden means of persuasion, financial or hallucinatory, the state geologist certified the east bank of the creek to be the foot of the Sierra Nevada mountains, based on its being a different color than the west bank. Aware of the farce but eager to see the railroad built because of the Civil War, President Abraham Lincoln put tongue in cheek and approved the beginning of the CP's triple subsidy for mountain construction.

Today in Old Sacramento, between Interstate 5 and the river, you can walk down the same streets as Judah and The Big Four, see where they drove the first spike, visit an HO model railroad of Donner Pass, ride a steam train, and tour the California State Railroad Museum to see the greatest artifacts of Donner and Southern Pacific history. The Museum is open daily and the ideal place to begin a visit. Your admission ticket also allows you to visit the reconstructed CP passenger station along the waterfront, where you'll find more rolling stock. The planned railroad technology facility is located about a half-mile south along the river, with displays and workshops for restoring the Museum's collections.

Among the great Southern Pacific artifacts you'll find near Central Pacific No. 1, the *Gov. Stanford*, are Judah's compass and other personal artifacts, as well as Alfred A. Hart photographs of the early railroad construction period. In the center of the Museum stands 4-8-8-2 Cab-forward No. 4294, designed like nearly all SP steam power to fit the measure of Donner's tunnels and snowsheds.

SP diesel locomotives on display include the impressive tiger-striped DRS-6-6-15 Baldwin road-switcher No. 5208. Representing speed rather than lugging power is E9A No. 6051 in brilliant Daylight paint. Being restored in 1990 is "black widow" F7A No. 6402, a veteran of Donner and the numerous other SP summits. A 4000 h.p. Krauss-Maffei diesel-hydraulic will be restored to its original external appearance after years of camera car work for the SP's locomotive simulator. SP steam engines on the property include the 4-2-4 type *C.P. Huntington*, 2-6-0 No. 1771, Northwestern Pacific 4-6-0 No. 112, and 0-6-0 switcher No. 1269 built at Sacramento in 1921. Donner rotary snowplow MW210 stands outside.

The Museum Library and its extensive collections are found upstairs in the reconstructed Huntington-Hopkins Hardware Store.

City of San Francisco leaves Sacramento westbound in 1967— Dick Dorn

SHIRLEY BURMAN

Ghosts of the Central Pacific's founders still seem to lurk around the large Sacramento Locomotive Works. Under the leadership of Master Mechanic Andrew J. Stevens, the Sacramento Shops became the greatest railroad repair and construction facility west of the Mississippi. A total of 195 steam locomotives were built here, from the experimental 4-10-0 *El Gobernador* to the SP's highly respected 4-8-2 and 0-8-0 types. In the modern era, the shops have reconstructed and upgraded more than 500 diesel locomotives. (Left) A winter's afternoon in the erecting shop finds SD40-2 No. 8530 getting its hood lowered after receiving a prime mover transplant. (Right) Out in the modern paint shop, Simone Morales uses 21 years of experience to apply SP lettering to the nose of the latest SD45 graduate of the General Rehabilitation and Improvements Program (GRIP).

In Old Sacramento, Theodore D. Judah looks toward the waterfront where his transcontinental railroad began. The Huntington-Hopkins Hardware Store provides offices for the California State Railroad Museum, where Chinese powdermen stand in a timeless tableau.

SHIRLEY BURMAN

SHIRLEY BURMAN

Inside the California State Railroad Museum are found the most sacred icons and artifacts of the Southern Pacific and its Central Pacific predecessor, including engineer Judah's compass and the road's first locomotive, *Governor Stanford*. That greatest artifact, Cab-forward 4294, stands ready to pull the first section of the San Francisco Overland Limited westward again across the Sierra Nevada barrier that inspired its design. Many an old-timer gets teary-eyed at the sight of this freshly painted locomotive— a million pound memento from many Western childhoods.

8816
8816
SOUTHERN
PACIFIC
ROSEVILLE

DICK DORN

Seventeen miles east from Sacramento is Roseville, western hub of SP's Central California mainlines and base of the 86-mile climb to the summit at Norden. As with steam locomotives, new diesel generations arise to conquer the snowy Sierra Nevada only to fall in favor of newer and more powerful locomotives. (Above) Dick Dorn pictures two of the vanished generations that once expedited freight movements eastward out of Roseville. No. 6389 and her 770 F-unit sisters dominated the Hill for two decades. Not as successful in the 1960's were the twenty-one 4000 horsepower Krauss-Maffei diesel-hydraulic locomotives like 9021. Despite great expectations, their maintenance-intensive high speed diesel prime movers and unusual design made them unsuitable for the demands of Sierra railroading. (Left) At the end of the 1980's, the axe is falling on workhorse SD45's like 22-year-old 8816.

Run 8 to Norden

"I never get tired of running over the mountain"

The safe and rapid passage of a typical freight train across Donner summit calls for sophisticated teamwork between road and helper engineers. Road foreman of engines, Todd Jacques, tells you how they do it.

"First, before leaving Roseville, you inspect your four six-axle units inside and out, making sure they have dynamic braking. You'll be needing plenty of it.

"After the train air brake test is successfully completed and the hand brakes released, you get a radio highball from the yard herder. Since you don't know yet where the slack is in your 5,500-ton train, you set between 5 and 20 pounds of air in your independent brake and notch out the throttle to Run 2. Your rear end helper engineer, with two units, is ready to go along for the ride so you ease off the engine brakes until you barely start to creep. You feel the slight resistance as each car responds to your pull, and a bigger tug as your helpers start rolling. Then you get a quick confirming, 'Helper moving', on the radio. Now release a little air to bring you to 5 miles an hour, controlling speed with the throttle.

"Your rear end trainman or helper engineer sings out, 'Mainline', as he clears the yard. Your entire train is on the high rail now and you tell him as soon as you spot a green signal ahead. He yells, 'Here I come', and starts shoving. You keep on building up speed gradually by taking a throttle notch each time your motor amperage peaks.

"Now you're around the big curve at I-80, headed into the grasslands of the Sierra foothills with everybody in Run 8.

"Flying over the No. 1 track at Rocklin at track speed, you hit the 1.5% ruling grade of the Harriman cutoff with its 50 mile an hour limit to Colfax. Between tunnels 16 and 17 your speed settles around 31 miles an hour, which means you'll be running about 21 miles an hour on the grades below Gold Run and Blue Canyon. Knock off a few more miles per hour for long trains or those with a lot of long cars, both of which create more rolling resistance on curves.

"Aside from calling detectors, watching signals, looking back at your train, listening to the engine, watching gauges and blowing the whistle, there's nothing else to do. To slow for signals or crossovers, you control your speed with the aid of your helper engineer, who notches back or adds power to help you out.

"Approaching Colfax, you slow to 30 miles an hour for a dip in the grade. You and your helper engineer really have to work together here. You've already told him you'll use a little dynamic today, so he knows to keep shoving over the top of the apex to keep the slack bunched before throttling back for the descent. You come out of dynamic around Long Ravine bridge and start carefully notching out again. You're both back in Run 8 with the roar of the engines echoing off the rocky face of Cape Horn as you look down a thousand feet to the American River. Hey, this is the way to run a mountain railroad!

"Historic old names flash past. . Cape Horn...Magra...and Secretown adjacent to I-80... before the short twisting 2% Gold Run grade begins, where speed drops to around 20 miles an hour with plenty of sand flying.

"There's a railfan taking your picture at the Gold Run crossing and you notch back a bit for the flat here, to stay at 30 miles an hour. You're in "the company notch" again as you pass Dutch Flat, where Dr. Daniel Strong once had his drugstore. He was a real mountaineer and the first person to tell Theodore Judah about a certain set of ridges that he thought might carry a railroad to the top of Donner Pass. "The weekend crowd at Alta store gives you a wave as you blast over the grade crossing and head for another view of the American River by Casa Loma campground, where the river is two thousand feet below. You hit the steepest part of the grade below Blue Canyon as the speedometer backs down again to around 20 miles an hour for the 2.3% incline.

"Grades don't last forever, so you're back to 25 miles an hour as you approach Emigrant Gap. The huge Bear River valley appears on the fireman's side as you pass the first CTC crossover and keep on grinding up toward the tunnels at Yuba Pass by the Highway 20 interchange.

"You drop back to Run 6 briefly and then slug it out at 25 miles an hour through Shed 10 on to Norden where the dispatcher has you lined you for tunnel 41, 'the Big Hole'. Your rear end helper, or FRED device, tells you you've got 90 pounds of air in the

The broad impact of SP's Sacramento upgrading programs can be seen at Roseville as rebuilt SD45 No. 7425 pulls out of the yard with the Roseville-Ogden Trailers, passing a pair of SD7's upgraded in 1978-80. The 1504 and 42 other "Baby Cadillacs" were successfully rebuilt for switching service, without dynamic braking or electrical transitions. Right, SD9 "Cadillacs" provide power for the Truckee Local. The 144 units of this class were upgraded electrically in the 1970-80 period. These six-motor units got their "Cadillac" nickname from crewmen who appreciated their smooth ride and dependability.

trainline before you thunder through the dark Norden sheds. Trains with over 100 tons per operative brake make a standing or running air test.

Topping the summit, you start notching back for the slow transition from full power to full dynamic. Full dynamic keeps you below 30 in the two-mile tunnel so you don't have to touch your train air brakes.

"Its full dynamic with a little air as you approach Shed 47 above Donner Lake. The same formula gets you down to Stanford Flats where you release your air and reduce dynamic braking to keep your speed up. Your helper engineer, who had been assisting on the steeper parts of the descent with a lot of dynamic, now is even pushing a little to keep you bunched through this lesser grade in Cold Stream Canyon.

"By the time you drop down through the Truckee crossovers you're working 200 to 400 amps of dynamic and done using brakes. Because of the undulating 1% downgrade of the upper Truckee Canyon ahead, your helper will shove most of the way to Sparks to control the slack, and help you control train speed by jockeying his throttle settings. You're doing your part, too, by adjusting your dynamic brake to occasional changes of grade.

"Finally, a few minutes past 6 hours and 155 miles after departing Roseville, you're whining down across Virginia Street in Reno and setting 6 pounds of air to come to a gentle stop on the mainline at Sparks.

" It was a normal run. No problems, mainly because a good engineer on the rear end helped make you look like a hero."

The whining descent of X6321 West down through Loomis with an Ogden reefer block typifies the early 1950's era when F7's took over Donner Pass freight service. Snowsheds were still mostly made of wood and open train order offices could be found at Newcastle, Colfax, Emigrant Gap, Norden and Truckee. (Right) By May 1984, orange colored fruit blocks have given way to yellow 9000-ton grain trains like this Ogden-Fresno unit train dropping down into Rocklin behind a trio of SD45's, with six more big units back in the train.

3199
SP

3708
3708
UNION PACIFIC
3708
3708
374
374

DICK DORN

(Above) The brown grassy hills above Roseville can still bring forth a call for that venerable Donner institution— the Fire Train. The SD9's of today's "Water Cars" from Colfax and Roseville meet below Auburn as hardy crewmen mop up the fire with high pressure hoses. (Left) Choice of locomotives depends on power and crew availability, as above Rocklin where a train led by tunnel motors and SD7's meet as they mop up a blaze. (Opposite page) Below Auburn, historic Bloomer Cut echoes to the passage of today's westbound *California Zephyr*, watched from atop the hill at right by Dalbert Castro, a Maidu Indian artist through whose ancestral ceremonial grounds the cut was blasted in 1864. Somewhere in Nevada, the UP SD40-2 was called to take No. 5 across the rails of its transcontinental competitor.

GOLD COUNTRY
FAIR
SEPT. 4·5·6·7
AUBURN

(Left) After flying across Interstate 80 on a high steel bridge of the 1.5% eastbound grade, Amtrak passengers get only a quick glance at the town of Auburn and the old Placer County courthouse. Unlike steam days, today's *Zephyr* seems to almost "fly" up the mountain grade—except at Colfax (Above) where it momentarily divides the old town with a unique stainless steel wall. The eastbound run of No. 6 from Roseville to Reno is about 15 minutes faster than the *City of San Francisco* in the 1950's. (Right) Soon after leaving Colfax, eastbound passengers draw in their breath as they round historic Cape Horn, more than a thousand feet above the American River. On these rocky bluffs the Chinese construction crews conquered their first great obstacle.

In a Sierra filled with gigantic unbridgeable canyons, the genius of Theodore Judah's use of a set of ridges to reach the Sierra summit is illustrated by the easy passage of this Roseville-Ogden freight around Rocky Point above Alta. Thousands of feet below is the rushing American River, yet the wooded ridge to which the train clings will carry it to Emigrant Gap and the granite ridges that lead to the 7000-foot summit.

(Right) An example of diesel railroading on the Emigrant Gap ridge in 1951 is Baldwin AS-616 No. 5235 rolling a westbound local down past old Nyak Lodge. The unit is scarcely a year old yet already deemed too slow for mainline helper service. (Below) Nearby in 1965, Donald Sims shot this view of the first locomotive to seriously challenge the first generation F-units. These two brand new 2750 horsepower Alco Century 628's carry the horsepower of nearly four F7 units.

DONALD SIMS

(Left) At Yuba Pass more than 30 years later, the lead "tunnel motor" of today's Chicago-Oakland Trailers rolls past the spot where the lead Alco PA of the *City of San Francisco* stalled in the snowdrifts of 1952. Drifts grew as deep as 25 feet along this side of Smart Ridge, and it was three days before the passengers could be rescued. (Above) Though six-motor units dominate Donner freight operations, we see GP40-2 and B36-7 power leading a grain train westbound past the same location, assisted further back in the train by a six-motor helper set.

The motive power history of Donner Pass shows many losers and only a few great winners— such as the dependable F-units of the 1940's and the powerful SD40 and SD45 models of the last 25 years.

(Above) We find the 6378 grinding up the hill past Troy in 1962, doggedly dragging tonnage toward the summit. (Right) In 1986, five big units whip an eastbound manifest up through Soda Springs on the last leg to Norden with the horsepower equivalent of ten F7 units.

9364
9364
SP

X6263

The Norden Experience

The westbound freight (Left) rolling out of the wooden snowsheds at Norden in 1951 represents a view of the Sacramento Division's mountaintop station virtually unchanged from the days of steam. It was located at 6880 feet above sea level a mile west of the Sierra ridge on what is probably the snowiest piece of mainline railroad in the United States,

Norden was a unique way of life for employees and visitors from its birth in the age of Sierra doubletracking, on March 10, 1926, until its end in September 6, 1985, in the age of microwave communications.

For railroaders, living at Norden was Heaven or Hell, and seldom anything in between. The beautiful mountain atmosphere with its constantly changing weather patterns and small-town way of life in yellow wooden company houses was paradise to those who chose to live in "God's country" and avoid the distractions of big city life. Nearly every railroad craft was represented among its residents, including operating, mechanical, signaling, track maintenance, bridge and building, water service, telegraphy and even food service. The nearest shopping center was the one room Norden Store and Post Office, up across old U.S. 40.

No automobile ever reached Norden in winter. If you wanted to "go to town" it meant getting on half the clothes you own and huffing and puffing up the hill inside a steep, icy pedestrian snowshed, to finally emerge at the headhouse near old U.S. 40. The few areas for parking were usually buried beneath the winter snowpack. Many employees, discovering their cars buried to the roof, would turn around and wait to catch a ride on the *Pacific Limited* or other train willing to pause for passengers in the dark Norden sheds.

Because of the nearly 7000-foot elevation, rare was the tenderfoot who didn't see a few stars and pinwheels as he neared a heart attack in trying to make this climb up to the highway after a heavy meal prepared by Fong Quong at the company cookhouse.

If there was a center of town it was that SP cookhouse, conveniently located between passing trains on the No. 1 and No. 2 tracks, where on and off-duty employees, family members and a few knowledgeable skiers would rub elbows beneath the warning sign saying train and engine crews will be served first. The windows of the building usually lay beneath the snowpack, except for those on the south side conveniently overlooking the sheds of the No. 2 track, where the flickering lights of passing Cab-forwards and passenger coaches could be seen through cracks in the shed timbers. In the diesel age there would be a gap between the sheds so diners could look directly out the window at the passing parade of first generation F-units.

Naturally, a few people considered their life at Norden worse than a prison sentence to another galaxy. They felt besieged in a world where snowstorms rage in June, rain falls every month, wolves and bears roam a granite landscape, coyotes feast on household pets, employee automobiles regularly get wrecked by plows, and where every six months the entire community falls into primeval darkness under the prodigious snows of a past ice age.

The SP's "Snowball Special" ski trains to Norden and the adjacent Sugar Bowl resort didn't resume after World War II. As diesel locomotives took over Donner Pass, Norden began a slow decline. The always-ready fire train went away, along with all the craftspeople whose work was tied to steam technology. The depot and most of the wooden sheds pictured at the start of this chapter went too, in a 1961 fire started by a foreman's overheated Long Johns.

But even in its latter-day role as a residence for track and signal maintenance people and telegraph operators, Norden cast its unique spell on a generation of employees and railroad fans.

The Rotary Chaser

One of those persons in whose heart the now-vanished Norden will always live is Dick Dorn, who with his wife, Chris, is a school teacher in Yuba City. Here he tells how his love affair began with this granite area of the Sierra summit:

"I can still remember my first ride over Donner Pass when I was ten years old. We were on a cross country trip on Highway 40 and I can still vividly see a freight train with a Cab-forward on the front and another back in the train as a helper. As a teenager I remember my first sight of a rotary plow at Roseville, vowing someday to photograph one in operation.

"Living in Marysville around 1972, with my first full time job, a new car, and some money, photographing Donner Pass became my number one priority. As a person who has always loved nature, the top of the pass was wonderful. When the weather was good in summer, it was fun to hike into remote and scenic locations to shoot the SP diesel operations.

"In 1974, I had a new little Shepherd mix puppy named Tarke—for the Sacramento Northern branchline out of Yuba City. We'd go camping in my yellow and white 1971 Volkswagen van. She and I made many trips together to Norden, often in winter snow conditions.

"This is about the time I met Albert 'Shorty' Teal, the second trick telegraph operator at Norden. We became very good friends. Many days were spent at Norden with Shorty telling stories of his lifelong railroad career. During the winter, he'd let me know when flangers and other snowfighting equipment were moving on the mountain.

"Though one of my big goals was to photograph the elusive rotary snowplows I never seemed to get the chance, even in the winter of 1975 when a heavy snowpack was left on the summit. I went up the Feather River Canyon and caught the Western Pacific's spreaders operating in the four feet of snow around Keddie. Later in the weekend I tried to get into Norden but found that the old road off I-80 was completely blocked by skiers who had run off the road and into each other. After several hours of waiting I got free from the traffic jam and headed for home. On the way I checked out Roseville and found the rotaries gone—obviously back up on the mountain!

Ever since that day I've had a dislike for the inept driving of flatland skiers.

"But a month later, Shorty called to tell me the rotaries were being assembled in sets again at Roseville and would be heading up the mountain the first thing in the morning! Music to my ears!

"The terrible storm scared most drivers away from the mountain that night, so I was actually able to get into Norden and stay over with Shorty and his wife in the basement they had fixed up for visitors.

"The next day was truly spectacular! With three flangers, a spreader set, and the rotaries all running— how could I miss?

"Stepping out of the Norden sheds in early morning as the snow let up, I shot the spreaders, and then a pair of flangers powered by SD9's. The rotaries showed up later, stopping inside the cavernous Norden sheds so the crew could go to beans. Their next job was to plow east on the No. 1 track which had not seen traffic for 24 hours and was covered by about 3 feet of fresh snow.

"Here was a chance to photograph them in deep snow working hard. I hiked, or really waded, up through the deep snow outside the sheds, and backed up against a vertical cut.

"I figured I'd be safe because they would be throwing snow away from the rock wall. But this was inaccurate, because they came out of the sheds throwing snow my direction! I knew I was far enough from the tracks to be safe, and I clicked away and got some great shots. As it passed, the snow thrown against the cut came down and buried me up to my shoulders.

"Then the rotary stopped and extended its wings, which scared the heck out of me. I was far enough away to be safe, but I'd learned a big safety lesson by the time I stepped out of the snow and realized how potentially dangerous it is to be around snow fighting equipment.

"The rest of the day was very productive and I finally left the mountain with a very satisfied feeling— having finally photographed the rotaries in action.

"Tarke was my constant companion on these Sierra trips. She would sleep next to me in the van. She was able to do fine without special equipment, but I needed snowshoes, down clothing and insulated boots just to survive. The Sierra snow is so soft and deep that it's very tough just to get anywhere, and when gale force winds and blowing snow come up, it gets much tougher! I realized, too, that it was always important to tell a friend where I was going in case of emergency.

"Tarke had a close call on one long hike with Ted and Liz Benson to Shed 10 in deep snow on a sunny winter day. It was tough hiking to get there and we were setting up to shoot the *Reno Fun Train* with a brace of F-units. But the eastbound *San Francisco Zephyr* arrived about the same time and I'd forgotten about Tarke as I climbed up in the deep snow to get a good vantage point. She walked across the tracks to see some people who had followed us up the hill, and apparently was struck by the pilot plow of the *Fun Train*.

"The trains passed and I couldn't see her right away. They said she'd been hit by the train. By the time I caught up with her she was wild-eyed but unharmed. Apparently the lead unit's plow had just scooped her off the rails and tossed her into soft snow. It was another safety lesson for all of us...she probably never heard the train that hit her.

"One of the other winter surprises on a trip with Tarke was to run into Dick Steinheimer at Norden, and deciding to hike together up to Cisco. I broke the way with my snowshoes and we went up on the hill over tunnel 3. Tarke and I scrambled to get ready for the westbound *Fun Train* as Stein set up behind us. I asked if he really wanted us in his picture, and he said, 'Yes'.

"Little did I know this picture would make Tarke a 'cover girl' and be used on the cover of the Northern California, *Growing Up With Trains II* book.

"After years of photographing the mountain in winter, I realized that the real beauty was often right after a storm ended. Those days, with the sun shining brightly and fresh snow covering everything like a layer of powdered sugar, were just made for outstanding photographs.

"Just a few years ago skiers got in the way again, at Shed 10 after I struggled up the hill with snowshoes and climbed a hillside with great effort to wait for the westbound Amtrak. It turned out I had inadvertently blazed a trail for a pack of cross-country skiers. They left ski tracks all over the places I'd been careful not to disturb. What can you do? Anyway, the photo turned out well despite the ski tracks you can notice upon close inspection.

"I think the most rewarding 'after the storm' luck I ever had was on December 23, 1982. Rotaries had been called for a big early storm and I was at Norden to catch them when the storm broke and the sun came out. It was rather unusual for rotaries to be operating under clear skies. Arriving at Norden at noon, they worked their way west under the bright sun, providing some unusual lighting for snow fighting

"Of all the memories of Donner Pass, none can compare with those of Norden. The little depot was built outside the north wall of the concrete shed so its doors opened into the shed where crossovers join the east and west mainlines. Inside, the operators controlled a CTC board they called an interlocking, which handled signals and switches between Truckee and Emigrant Gap.

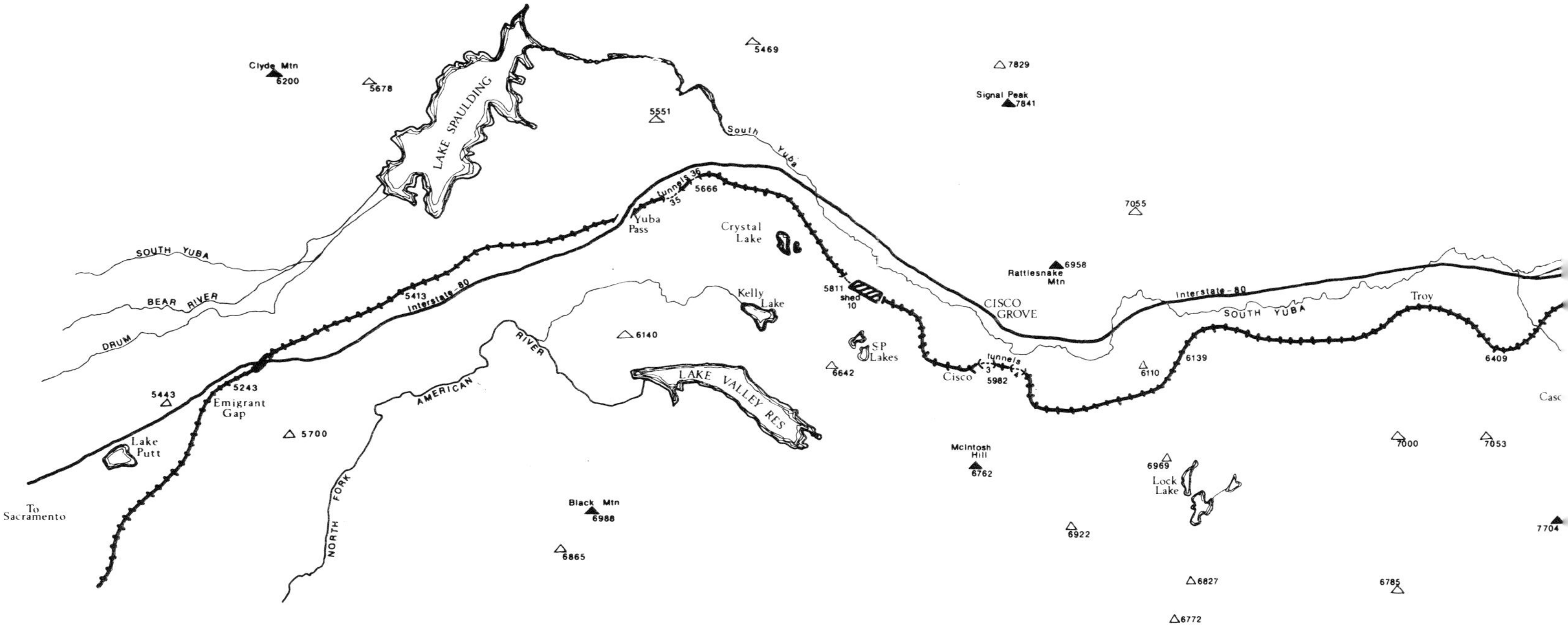

Emigrant Gap to Truckee 35 rail miles

1 mile

Shirley Burman Design 1986

"Connected to the depot by hundreds of feet of pedestrian snowshed 'chutes' were the employee residences, a two-story dormitory for snow crews and officials, a beanery and a number of equipment rooms. The whole area was usually off-limits to the general public and railfans, mainly because the employees didn't want the distraction and people could easily get hurt walking around in the dark sheds. Also, in winter, tempers tended to get short as the long season wore on.

"I had permission to be there so I got to know the operators and many of the officials who made Norden their headquarters when storms complicated operations. It was a real luxury to be able to sit inside a heated office in a comfortable chair and watch the progress of a train on the CTC panel so I could wait until the last minute to get out into the storm and find a place to shoot the train.

"In Norden's final years there was an interesting group of operators working there, three with the same sounding last name: Doris Teel, Danny Thiel and 'Shorty' Teal. The chief operator was Harry Hanns, who like Earle Stice before him, handled a great number of emergency situations. It was not unusual in winter for several lost skiers or hikers to call for help from one of the block phones. Mary Hanns was postmistress at Norden store.

"The day after Norden closed, Doris took me down to the depot and we recreated some of the last day scenes for pictures. I was able to clean out the trash cans where copies of the last lineups and orders were still waiting to be disposed of. It was a real treat for me to save this last paperwork from a place I really loved. The weather outside was a dismal drizzle as I spent my last hours in the depot— weather that was very typical of Norden"

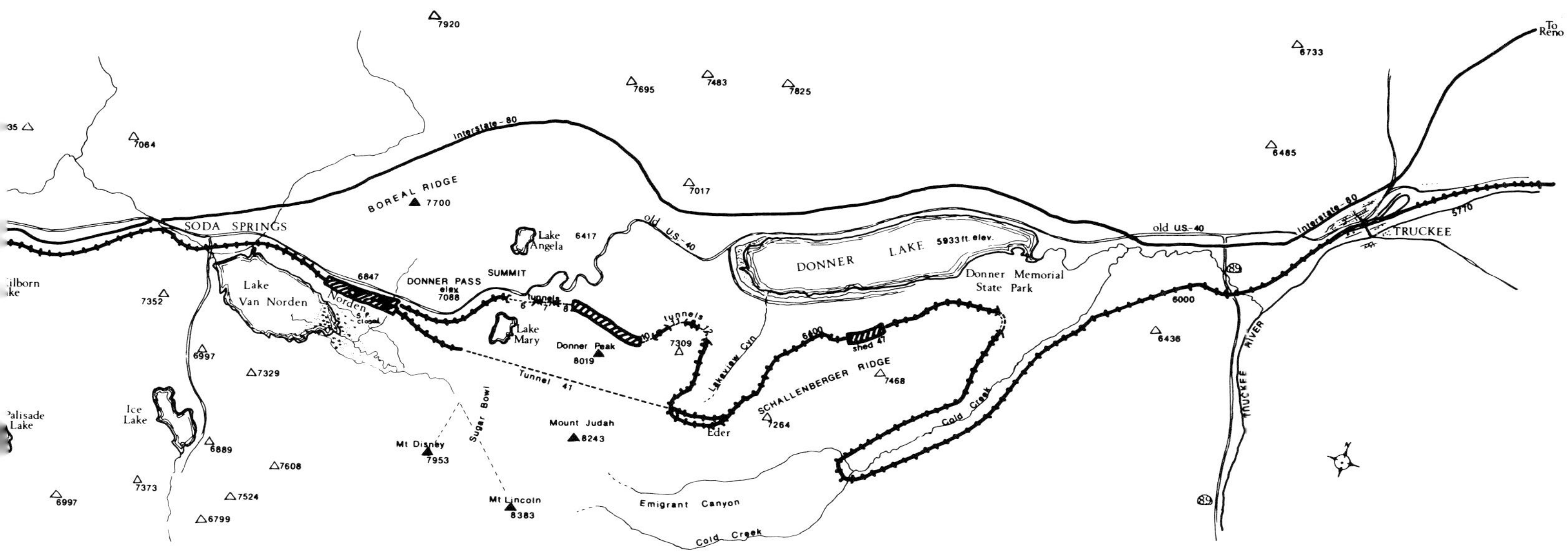

Southern Pacific Railroad Over Donner Pass

The Norden Weatherman

A modern visitor who was fascinated by Norden and shared his weather forecasts with SP railroaders is Michael Pechner, a trainfan who operates Golden West Meteorology forecasting service in Cordelia. As an avid follower of trains and weather, Mike had many of the same appreciations for the people and rotary snowplow operations as Dick Dorn, and also was a frequent guest of the Teals. In fact, he installed his own phone there to call in weather reports to radio station KCBS in San Francisco. He now calls in to station KNBR from the bed and breakfast lodge at Norden House:

"Charlie Crocker and his workmen wouldn't have believed it, but the first 15 or so winters of the Central Pacific were fairly near the normal snowfall of 450 inches. That era ended in the winter of 1879-80 when 790 inches of snow fell on the summit, again convincing everybody of the value of wooden snowsheds. But it fell fairly evenly throughout that winter, making it easier to handle than what happened 10 years later.

"The decade of the 1890's was ushered in by a 780-inch snow season, with most of it falling in just a three month period. This 1889-90 season was the grandfather of storms at Truckee and Blue Canon (pronounced.Canyon) and resulted in the railroad being closed down for several weeks. A long stretch of sheds burned down the previous summer above Cisco and had not been rebuilt. The only tool that proved able to reopen the railroad as quickly as it did was the CP's first Leslie Rotary Snowplow. When you look at books and travel around in the snow country, most of the old time pictures you'll see of heavy snows are from this memorable winter.

"The success of the Leslie machine in this stormy period led to SP eventually owning 15 rotary plows, and to a great extent to the decisions to keep cutting back the number of miles of snowsheds across the summit from a high point of 38 miles.

"Most people have the vision of SP running rotary snowplows every winter, but that's hardly the case.

"They were out in full force for the winter of 1937-38 in which Bill Hack kept his railroad open using everything from Cab-forwards to 2-8-0's. The season's total snowfall at Norden: 68 feet and 3 inches, the heaviest Sierra winter ever recorded. Hack was considered a 'superintendent's superintendent' by the people with whom he worked.

"He was gone by 1945 and the famous winter of 1951-52 proved to the new management that there are times when the line should be shut down before problems multiply, as witnessed by the stalling of the *City of San Francisco* at Yuba Pass. The culprit was a great amount of snow falling in a short time, coupled with high winds which created gigantic unstable drifts along Smart Ridge between Crystal Lake and Emigrant Gap. This time the railroad was shut down for a week and U.S. 40 for five weeks.

"Winters through the rest of the 1950's were pretty quiet, and the rotaries didn't venture out very much, in fact they weren't seen through the early and mid-1960's. Then in the winter of 1966-67 over 500 inches fell at Norden, with 170 inches on the ground at the end of February. The spreaders worked with the rotaries westward from Norden by spreading the No. 2 track and pushing excess snow onto the No. 1 track for the following rotary to heave over the bank, a common technique used today for clean up after storms.

"The rotaries were out again in the 596-inch winter of 1968-69 which resulted in a near record pack of 22 feet and 5 inches on the ground at Norden on March 1. Rotaries came out again in the winter of 1974-75.

"More recently the greatest use of rotaries was in the winter of 1981-82 when heavy snow and slides caused a 5-day blockade of the hill until things could be brought safely under control. The next year was the second greatest snowfall in history, with the railroad staying open despite a season total of 796 inches. Because the snow tended to fall more evenly through the season, rotary use was minimal.

"Though high horsepower tunnel motors are standard power for rotaries, even today the flanger and spreader crews prefer the rebuilt SD9 'Cadillacs' over other power, because of their smooth ride, lugging ability, braking power and overall dependability.

"Remember to be very careful in these mountains because heavy snow can come at any time. On September 6, 1986, two feet of new snow fell at Norden. Almost as much fell on June 20, 1975, the day after the passage of Daylight engine 4449.

"In case you don't think dry periods can be dangerous, note that 52 inches of snow fell at Norden on February 3, 1989.

"I've been lucky in taking pictures on the mountain because of being a meteorologist, having advance knowledge of the storms and the chance to associate with the Southern Pacific operators and officers who used to work at Norden during the snow season.

All of those people are gone now from Norden, retired or working in other places on the railroad. A few have even risen to the top levels of SP management. I think the majority of these railroaders would say that handling winter operations at Norden was one of the most memorable challenges of their career.

They were great people."

Woman on the Mountain

Shirley Burman got her first taste of Norden in the snowy spring of 1983. From a woman's viewpoint, she describes the fun and difficulties of photographing on Donner:

"Like most travelers on I-80 and old Highway 40, I didn't realize that there was a little cosmos of railroad workers and their families living along tracks and sheds in the Sierra at Norden. Only railroaders and railfans seemed to know about the 'the chute people' I met on my first trip there on April 30, 1983. The 'chutes' were covered wooden passageways needed in winter when deep snows made it impossible to walk between the railroad buildings.

"Steinheimer heard that the depot roof was caving in from the weight of the snow and he thought it might give us the chance to do some interesting spring picture taking. We grabbed our cameras and away we went to this town where the snow was still about 12 feet deep. I had never seen so much white before. It was almost impossible to walk in, so I let Dick break the trail through the deep drifts. Mistake!!!!! Anyone who knows him knows that his 6' 6" height is made up of two legs and a head, which gives him a heck of a stride. So I spent most of my time climbing in and out of his hip-deep foot steps, struggling to keep my camera lens out of the snow. I looked like a snowcone maker with my camera swinging out from my neck, scooping snow as I struggled through the drifts.

"One of the side effects of hiking at Donner Pass is the feeling that at any moment you're going to have a heart attack and/or keel over from lack of oxygen. I puffed and huffed to where the depot was supposed to be. But where was it? All I saw was a buried concrete snowshed and railroad men in yellow slickers with shovels scurrying around moving snow from one spot to another. I asked, 'Where is the depot?', and Dick said, 'You're standing on it.' Then a real flurry of activity began, more snow was starting to fall and the heavy snow on the roof was getting to the critical stage. There were not enough men to shovel the snow fast enough, so SP's Charlie Drinnon called for a tractor to push the snow off the roof. It was quite a sight so I took a lot of pictures of all the activities, with the nearby section houses almost buried by the deep snow with just their little roof peaks and chimneys sticking out.

"While standing on this deep snow pack, people would disappear, and then reappear. I walked around to see where they were going, and it turned out to be a straight down tunnel through the snow with a ladder to get to the depot door. Later in the summer I went back and couldn't believe that a few months earlier I had been walking over the top of chutes I had never seen, even over treetops and rocks. The only reason I'd seen the back of the houses was because they were two stories high.

"When it was finally time to make our way back up the chutes to the highway, I found that maneuvering over the frozen wood floors and up the steps took a lot of energy at that 7,000-foot elevation. For a flatlander, it was a real effort just to breathe. I groped up through the darkness hanging onto the railing with only an occasional bare light bulb to light the way. My experiences of hiking through those cold and dark passageways still give me the shivers.

"The next couple of years we spent photographing the Norden folks doing everything from making tea to mowing their lawns. These 'chute people' were soon to be history and we didn't want to miss a thing. People weren't always my focus of attention, like any other railfan I do my share of train shooting.

"I never thought I'd get excited hearing the throbbing of an approaching train, especially a diesel, but it happens to us all. I remember the first time walking through the dark tunnels at Norden and seeing the glow of a locomotive headlight. What a sight. The outline of the square sheds in blackness and the silhouetted black engine with nose glowing makes the heart pound and makes a great picture.

"On another quick trip to Norden, we heard that snow and ice was creating havoc in the sheds. The snow had melted between the rails and frozen up, causing several locomotives to derail. I was attempting to take pictures in the cold and darkness and finally gave up after almost getting frost-bit fingers. Besides, it was time to check out the warm cookhouse. Seeing three men eating at the counter, I thought it might make a good picture. The guys continued eating with an occasional scowl turned in my direction, but they did seem to be tolerating my picture taking. Then to my horror I noticed I had no film in the camera just as they finished and were getting ready to leave. I knew I didn't want to miss this chance of getting their pictures, so I swallowed my professional pride and asked if they wouldn't sit down for just a moment and pretend they were drinking coffee or something. I figured what the heck, the most they could say was 'bug off' and get up and leave ..... or they might just cooperate and let me have another crack at taking their picture. The three men looked at each other and reluctantly agreed. I snapped off a half-dozen shots, they left and I hoped they wouldn't see me again. Then Dick tells me, 'You just photographed the big brass, one of those guys was the toughest boss on the mountain, Bill Lynch, the division Chief Engineer." After that whenever I'd see Lynch, with a mock scowl he'd ask if I had any film in the camera this time.

"Donner Pass wasn't always snow, cold and misery. Dick and I made many trips to the mountain in spring and summer. While waiting for a freight or Amtrak to roll by, we'd see wild flowers in bloom, deer running through the brush and skipping over the rails. I got to hear the rushing streams catapulting their way over rocks and through crevices while racing to Donner lake. We hiked the rails above Truckee with Dick telling me extraordinary old stories about the building of the railroad and about contemporary times. We'd get all charged up when we knew 'something was coming' and it would turn out to be an unexpected unit train, or clean locomotives. On one trip we hiked through old wooden snowsheds, and I heard the rushing cascades of melting snows sliding down the side of the sheds and felt the drips from the dilapidated wooden roof that turned the inside of the sheds into a rain forest.

When we entered a tunnel, Dick said, 'Hush! .... listen to the falling water, you can hear the voices of the construction workers from more than a hundred years ago.' He was right, I did hear the voices. I even got to see the same tree above tunnel 8 that was in Alfred Hart's picture of the Chinese tea carrier.

"Truckee was always an important destination for us, because we'd look up Charlie Drinnon and get the scoop on everything going on. While Dick was getting the low-down on train movements, I'd try to photograph Charlie, one of my favorite railroad people. He's photogenic with his ruggedly handsome tanned face (I would notice those kinds of things) but he didn't believe in cooperating with people with cameras in their hands so he'd keep on the move. It was a constant challenge for me to outrun him. On a few rare moments we'd find Charlie in his office in the old Truckee depot, leaning back in his chair with his feet up on his desk puffing on a cigarette, the phone would ring.....I'd do a grab shot and the race would be on again.

"Our many hikes were invigorating and fun, but my all time favorite trip was on one assignment for SP when we high-railed from Sparks to Emigrant Gap with Jim Mahon. That's the only way to go and see it all, unless you're a masochist.

"I've only had a brief romance with the mountain and Norden, but it was wonderful while it lasted. Norden was bulldozed down a couple of years ago and the bushes and flowers are taking over again like Norden had never existed. The wooden sheds are mostly gone too. The Norden operators have been replaced by a computer dispatching system at Roseville. I can't even find Charlie Drinnon with his feet up on his desk in the old depot anymore. The (Truckee) depot was given to the city and is now beautifully restored and utilized by Amtrak and the Chamber of Commerce. Charlie and the mountain are still around though, just in different settings."

SHIRLEY BURMAN

Life at Norden in summer was a combination of outdoor beauty, and interior darkness within the snowsheds and interconnecting pedestrian walkways, such as the steep tilted stairway leading up from the depot to the snow-measuring table and residences.

(Above) The old saying you could follow a steam engine through the sheds by its smoke still applies to diesels at Norden. (Below) Extra 7326 West pops out of the asymmetrical structure of Shed 26 1/4.

(Left) Sometimes called "The Mole People" because of their subterranean life, Norden residents often found it hard to tell spring from winter. This year it's May 12 before operator Sandy Strawn can finally get outside to clean her car. (Below) The same scene on April 30,1983, still finds 12-foot drifts standing from the season's 796-inch snowfall.

Buried beneath the Sierra snowdrifts in an environment almost as remote as a space station, operator Louis Stearns Jr. waits out a quiet night alone at the Norden interlocking board. "How can you possibly LIVE there?" his Florida relatives once asked.

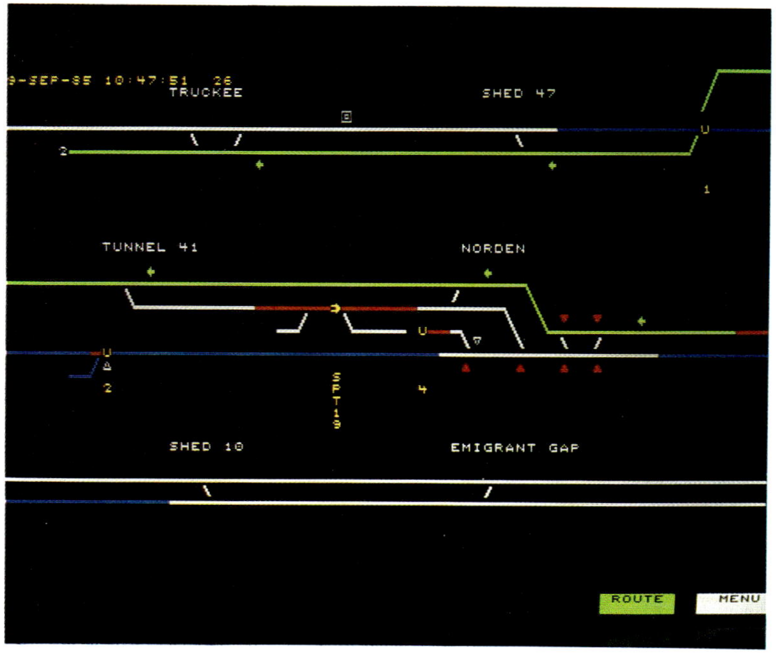

Microwave technology finally brought an end to 60 years of life at Norden on September 6, 1985, where (Left) a section of wooden sheds still remains to protect trains like the eastbound *California Zephyr.* (Above) Dispatching between Emigrant Gap and Truckee is now done at Roseville on this CRT display. (Right) On the last day, operator Doris Teel stayed on duty to assist a Roseville dispatcher with the new CTC system until passage of the eastbound Amtrak. Then she sadly walked home with her dog through the mists of the community's last Autumn.

3103
3030

(Left) SP and Rio Grande power leads a heavy train eastward from Norden on the No. 2 track toward the two-mile tunnel 41 beneath Mt. Judah. (Avove) The dispatcher has brought Extra 8569 East up to Norden on the No. 1 track and is holding it at the shed entrance to be overtaken by this morning's Amtrak No. 5 on the No. 2 track.

Grinding up the original CP line across the Sierra, this long westbound train stretches back through a mile of snowsheds and tunnels on the No. 1 track beneath Donner Peak. Behind and below the locomotives are walls of hand hewn granite blocks built by the Chinese stone masons during the original construction. Railroad builders and emigrants were tested to the limits of their endurance by this most difficult place of the Westward Movement.

Mountain Soul of a Fighting Railroad

"East of the summit, the tunnels and sheds of the No. 1 track on the side of Donner Peak stand as great monuments to the fighting spirit of the Central Pacific and to the Chinese in America. Here on this incredible mountain battlefield it took as many as 10,000 men three years to complete a railroad your eyes can see in a few seconds."

If the fighting soul of the Central Pacific wasn't formed in the early 1860's in Sacramento, then it was formed in the attack by Central Pacific forces on the hard granite faces of the Sierra above Emigrant Gap, and in two winters of men working and living beneath snowdrifts to complete the eight tunnels at the top of the pass.

Chinese work camps were carried away by avalanches at several points in the summit area, and often the only way to reach other camps in winter was by means of buried snow tunnels inside the drifts laying against the sides of the mountain.

Today, with electricity, internal combustion engines and huge construction equipment, its hard to imagine the building of this line in one of the toughest environments in the country by a work force living without proper shelter using nothing but hand tools, with only human or animal powered vehicles.

Hell has always been assumed to be hot, but the Central Pacific workmen certainly found it in the wet, blustery cold of tunnel No. 6 at the very top of Donner Pass.

By Judah's surveys, it was essential to pierce the quarter-mile granite summit ridge to begin the descent of the eastern slope toward the Truckee River and Nevada. The first excavations for this 1659-foot tunnel were made in August 1865 before heavy winter snows shut down all construction above Dutch Flat. Work was resumed again in the spring of 1866 and continued through the winter until the top of the pass was finally holed through in August 1867. Working four faces at once with hand drills and black powder or nitroglycerine, the total average progress through the hard granite was about three feet per day.

Most historical accounts imply that the worst of the Sierra battle ended with completion of the summit tunnel. Actually, there was still the laborious completion of seven more tunnels east of the summit before the line could even reach the Nevada border.

At Cheyenne in the winter of 1867-68, Jack Casement and the great construction army of the Union Pacific was poised to head west across the gently rolling hills of Wyoming to claim as many miles as possible of the future transcontinental railroad.

Time was of the essence in this race against the Union Pacific and the Central Pacific was bogged down by the same barriers that had trapped the Donner party. Progress toward Nevada this winter was again going to be measured in feet per day.

Thus the fate of the Central Pacific was hanging in the balance in late 1867 as Crocker and his foreman, James Strobridge, devised a daring plan to escape their granite trap.

Starting 13 miles back at the snowed-in railhead at Cisco, they gathered hundreds of teams of animals and probably thousands of men, to began sledding three locomotives, cars and several million pounds of track construction materials eastward up toward the summit through mud and snow. After struggling to the summit, they sledded these heavy supplies, which had already come 15,000 miles around Cape Horn, down over the same snowy precipices that had blocked the emigrants. As the heavy loads reached

Philip Choy of the Chinese Historical Society stands in an icy No. 1 track snowshed 125 years after men of his race were living beneath snowdrifts to blast this line out of the granite face of Donner Peak.

Donner Lake they were transferred to wagons for the rest of the trip to Truckee to begin construction of 35 miles of railroad down the Truckee Canyon toward Nevada.

This tremendous winter effort by men and animals was successful. The tracks were already complete to Reno by June 18, 1868, as the first through train finally made its way down off the Sierra grade into Truckee. In terms of sheer risk, uncertainty and struggle, the mile-a-day race across Nevada to Promontory was a school picnic compared to this perilous battle of the Sierra summit.

In later years when people asked Charles Crocker what he did, he only needed to say, "I built the Central Pacific!"

Because of the essential toughness of Crocker and his workforce, and the financial maneuvers of Huntington, the Southern Pacific still remains one of the very few major railroads which has never been bankrupt.

Today you can park you car near the summit of the old highway east of Norden and walk over to the south side of the road to see the square cement and steel platform which defines the center haul hole of summit tunnel 6. Here's where in April 1867, Crocker and Strobridge stood and argued with Leland Stanford. The latter was insisting they try out a new invention he and Hopkins had purchased and sent to the summit to speed up progress on drilling the tunnel.

Probably remembering their troubles using another new invention, nitroglycerine, Crocker and Strobridge refused to set the new machine in the shaft, telling Stanford that even if it worked it couldn't help because they had already reached the hoisting capacity of their only steam boiler.

So the Central Pacific was probably the last major construction job in North America done entirely by hand tools. The invention Crocker and Strobridge declined to test that day was the steam drill, predecessor of the pneumatic drills still used today.

Though most westbound trains now cross over at Shed 47 to use the two-mile tunnel route of the No. 2 track to Norden, a few westbound trains still climb the original grade of the No. 1 track along the shoulder of Donner Peak. As they pass through the tunnels and sheds and along the Chinese walls approaching tunnel 6 they are crossing what is probably the single most difficult place of passage in the history of the Westward Movement.

Building this line was the work of giants— one of the largest peacetime armies ever fielded in America. Most of the "giants" were short in stature— Chinese workers from California and those conscripted from their native land. In their hard labor and suffering they created the greatest monument to the Chinese in America. The race that built the Great Wall of China came into this new land to construct the most difficult portion of the West's first great transportation corridor.

Tunnel 6 at the summit of the No. 1 track— where winter still lives.

SHIRLEY BURMAN

7352

The westward climb of heavy trains up over the Chinese wall of the old No. 1 track over Donner Summit occasionally results in titanic battles between locomotives and the forces of gravity and friction. This morning's too-heavy coal and merchandise train shows the utmost resistance to movement as engineer Chuck Adams' three overheating six-motor units struggle out of tunnel 8 across the Chinese wall to the accompaniment of vast clouds of exhaust. A stop must be made, and then the battle is rejoined with engineer Lonnie Dixon's five big units battling their way across the wall and disappearing into summit tunnel 6. Could even the struggle of a half-dozen wood burning steam locomotives been more dramatic?

The idea that massive timber structures such as the No. 1 track snowsheds ahead have only a finite life is probably not being considered by the passengers of the westbound *Zephyr* as it climbs through the tunnels and sheds to the summit. (Right) The routine passage of another day's No. 5 through the old timber portion of Shed 37 leaves vivid memories of the dramatic "Star Wars" lighting effects of sun and diesel exhaust.

4305
4305

(Left) Two weeks after the wooden shed pictures on the preceding pages were made it became apparent that sheds do have a finite life. The great 12x12- and 16x16-inch timbers have already been pushed aside to begin the job of concreting this 500-foot portion of Shed 37. Surveyors work to lay out the footings and connections with the existing Chinese wall. By July, the first six-ton interlocking concrete side panels are being placed, immediately followed by placement of the six and one-half ton roof panels. (Right) It was wisely assumed by Bridge and Building foreman Dick Carter that the typically adverse Donner weather would play a hand in completion of the shed. So it is, in late October, when carpenter Mike Ivanusich hurries to bolt down the last roof panels before the full wrath of winter falls on this shoulder of Donner Peak. Two hundred miles west of the summit, San Franciscans swelter in a heat wave.

7504
SOUTHERN
PACIFIC
7504
7463
SOUTHERN
PACIFIC
7463

(Left) A throaty roar fills Lakeview canyon above Shed 47 as upgraded SD45's thunder across the east portal of two-mile tunnel 41 with a heavy Ogden-Fresno grain train. At this point Extra 7504 is crossing over to right-hand running after the 50 miles of left-hand operation on the No. 1 track west of Sparks. (Right) Only 14 years earlier the sounds along Schallenberger Ridge were those of ex-Union Pacific E9 units challenging the grade on a rainy day near tunnel 13 with Amtrak's then *San Francisco Zephyr*. These E units never gained the respect of the Alco PA units on the Donner grade.

The early summer sun at Stanford Curve falls on the SP's *City of San Francisco* with its mid-1960's "City of Everywhere" consist as it storms up the westbound grade above Truckee. Standard power was a trio of Alco PA's but when traffic was heavy it was common to add a single F7A unit, such as today, to provide extra power or to operate the Automatic Train Stop system between Truckee and Gold Run when that on the lead Alco unit was inoperative. The ATS was removed in 1968. (Right) In 1966, brand new six-motor SD40's are arriving from General Motors to work with remnants of the F-unit fleet, hauling such loads as these vintage refrigerator cars headed back to the irrigated fields of California.

SHIRLEY BURMAN

(Right) The name of the game on this Fourth of July afternoon in 1984 on Schallenberger Ridge is swatting mosquitos and waiting for an eastbound Fresno-Ogden manifest. Finally, as the sun sinks low in the sky, the train rolls grandly out of Shed 47 to the clicking of shutters. (Above) Then as it grows dark, the day is made complete by looking down on the beauty of a fireworks show on Donner Lake.

SHIRLEY BURMAN

9294 9294

4335
SP

(Left) The spring of 1984 finds five beautifully matched SD9E's dropping down into Truckee with the Nevada Cement empties headed for the company's facility near Wadsworth, Nevada. (Above) This occasion of all first generation power on a unit train is matched in rarity by this daylight passage along Schallenberger Ridge of the very hot Oakland-Chicago Forwarder. With today's expedited schedules after the Rio Grande merger, the midnight hour is now more typical for this train's Sierra crossing. This ridge is named for Moses Schallenberger, a teenager fallen ill and left behind by members of the Stevens Party of 1844. He recovered his health and became the first known person to survive a winter at Donner Lake. Two years later his cabin would be used by the Murphy family of the ill-fated Donner Party.

(Left) The Upper Truckee River Canyon below Truckee provides a welcome contrast to the ridge top climb of the west slope of the Sierra. Here one can look down on the beautiful ripples of this fly fishing stream and see the descent of a heavy forest products train on the gentle gradient.

(Right) The ascent of the Truckee River Canyon is not quite as easy as the descent, especially with the heavy loads of Nevada Cement added to this westbound train at Sparks. Generally, Donner trains are assigned about 1000 tons per six-motor unit. Even on the 2.3% west slope, this allows enough speed to prevent train handling problems.

SOUTHERN
PACIFIC
3204
Grande

DICK DORN

(Left) The heaviest regular train over Donner is the Mojave Coal, received from the Rio Grande at Ogden. Ten big units are often used, and on weekends in 1984 it is still common to find four or more SDP45 San Francisco-San Jose commute engines getting a snoot full of hard work on the westbound grade, such as here above Boca in the Truckee Canyon where the thermometer once hit 45 degrees below zero.

(Above) The gambling-oriented Reno, Nevada, skyline is the backdrop for today's westbound Elko-Newark salt empties, a completely improbable unit tank train operation which hauled salt brine from California's Bay Area to a chemical plant on the Great Salt Lake. . . temporarily in need of brine concentrate because of Lake flooding.

RENO
THE BIGGEST LITTLE CITY IN THE WORLD
Commercial Row
Fitzgeralds
WIN
BIG
MEGA BUCKS
QUARTER MANIA
NICKEL MANIA
SOUTHERN
PACIFIC
GEM

DICK DORN

Car and pedestrian traffic has come to a momentary halt in Virginia Street's "glitter gulch" as the rear end helper of a heavy Ogden- Fresno unit grain train clears the crossing with a loud roar. Inside Fitzgeralds Casino and nearby Harold's Club, the gamblers at the green felt tables and the slot machines pause for a fraction of a second for the unexpected roar of the 20-cylinder diesel blasting west toward the dark Sierra ridge. Then sounds of the big tunnel motor recede, and fantasy returns to downtown Reno.

SHIRLEY BURMAN

Sparks, at the eastern end of the Donner Grade, is where we find engineer Susan Bowlus and her rebuilt SD45 locomotive ready to switch additional cars to the rear of the westbound *Zephyr* for the benefit of weekend gamblers returning to the Bay Area. (Right) The five 3800 horsepower GP60's rolling down into town in mid-1989 are the first third-generation units to directly challenge the supremacy of SD40-T2 tunnel motors. The test results: Excellent four-axle locomotives, but heat from the tunnels and long grades caused horsepower deratings of more than 50%, making them less effective on Donner than six-axle tunnel motors.

9628
SSW

Winter's Professionals

"Want to experience the snow that trapped the Donner Party? No problem. Park your car, jump up on a fresh 10 foot snowpack and go for a nice long hike. . .say 30 or 40 feet."

Watching the snowfighters of Donner Pass like (Left) foreman Arcadio Ruvalcaba, is like watching professional linemen in a Super Bowl— those guys who get pounded into the mud every play so guys with clean jerseys can win the game. With these "Winter's Professionals", SP shippers are the guys with clean jerseys who get dependable trans-Sierra transportation. Occasionally even UP sends a few trains over Donner when rain or snow brings down slides in the Feather River canyon.

The first moves in the game take place in summer, with maintenance of way, bridge and building, and signal crews doing their summer repair work and preparing for winter. Sheds are rebuilt, new rails are laid and the two 70-ton Jordan spreaders are shopped.

Winter begins with arrival of the first storms of September and October, ending in April or May after an average annual snowfall of around 37 feet .

The first snows are generally light, only pencil notations on a dispatcher's train sheet. But superintendent Mike Irvine and his staff are paying close attention to winter forecasts, and have several SD9 or GP38-2 flanger sets ready at Roseville before the first good storm hits.

The duty of the flangers is to keep packed snow from accumulating between the rails where it can lift air hoses or locomotive fuel tanks. They aren't called until there's at least 15 inches of of snow on the rails.

The maintenance of way crews of Charlie Drinnon at Truckee and Ray Chavez at Colfax stay on alert, shoveling out any affected mainline switches. Between Emigrant Gap and Truckee, all CTC crossovers between the two mains are protected by concrete snowsheds— Shed 47 above Donner Lake, Norden at the top of the pass, Shed 10 below Cisco, and the I-80 underpass at the Gap.

Everything is based on keeping ahead of changing weather conditions. When forecasts say the storms will continue, another flanger rolls out of Roseville. With their locomotive pilot plow and flanger blades they can handle up to a three or four feet of snow as long as there's a place for the snow to go. By then, Drinnon is preparing to make a pass across the mountain with his double-ended spreader set, to clear snow from the right of way so the flangers can either return home or operate more efficiently.

When a series of storms starts piling up 3 to 10 feet of new snow spreader crews keep removing the "core" between the parallel mainlines and pushing snow away from the right of way.

Why this heavy use of spreaders rather than rotary plows?

The spreader is a super efficient mover of snow up to around five feet deep. When Drinnon's crew comes rolling down the track at 20 miles an hour with the arms spread to clear a 40 foot swath, shoving snow with a water content of 34%, every second they're moving about 125,000 pounds of snow from the rails. That's about 7.5 million pounds of snow a minute. No other machine can match that.

The four wide-wing SP rotaries kept at Roseville are still important because high winds combined with a heavy storm, or series of storms, can form snowdrifts 10 to 30 feet in depth. Spreading gets real tough in snow deeper than 10 feet, so this is where the rotaries really shine. They move in slowly, cutting a trench 16 feet wide, easily handling 12-foot drifts. Rotaries can handle even higher drifts by plowing in a short distance and then backing up, allowing the overhang to collapse before moving forward again. But in actual practice, bulldozers are already at work clearing off this excess height snow and pushing snow away from the rails to give a place for flangers and spreaders to toss more snow.

Like spreaders, rotaries are generally run in double-ended sets, with two or three big motors in between for tractive effort. That way crews can always fight their way back out of trouble.

Names of workmen from the past century like Toy, Lee and Ling are gone from the mountain in the 1990's, replaced by men with names like Jimenez, Martinez, Carrillo, Drinnon and Mahon. They work here because its what they want to do, taking care of business on a railroad that for 125 years has been the ultimate proving ground of men and machines.

DICK DORN

Very cold Gulf of Alaska storms bring snow to all altitudes of the Sierra Nevada, such as at Colfax, where a trio of Rio Grande units help ease the Chicago-Oakland Trailers down past the old SP station that once saw the little trains of the Nevada County Narrow Gauge Railroad.

(Right) At Truckee, a flanger crew running ahead of a Roseville-Ogden manifest has backed into the house track to allow it to roll past toward Sparks. Passage of the manifest across the mountain has been greatly helped by the snow hurled off the tracks by the SD9's pilot plow and by the flanger's removal of the snow buildup between the rails.

9231
9231
SP
4348

4319

(Left) As the fight against the 1983 storm continues, flanger power like 4319 in the sheds at Norden begins to look like a cake decorator's icy dream. Aboard, trainmaster Gerald Nester gives a smile before engine and flanger are heading off again toward Emigrant Gap to turn on the balloon track. (Right) SP is tight for motive power today, so the 120-foot turntable in the Norden sheds is used to turn the extra rear end helpers keeping trains moving in the storm, the last uses of the table before being damaged by the extremely heavy snowfall.

DICK DORN

As traffic spins out of control on Interstate 80, flangers keep "flipping" between the balloon tracks at Truckee and Emigrant Gap. Both GP38-2 No. 4841 and its flanger are moving plenty of wet, heavy snow. (Left) Though the tracks are clear, more storms are coming as this eastbound freight approaches the top of the grade in the sheds at Norden.

4438

(Left) Storms at Truckee find Charlie Drinnon's maintenance of way gangs clearing snow from a pair of SD9 units to be coupled to Jordan spreaders to widen out snowbanks on both mainlines. (Right) Like the SD9's, these snowfighting spreaders with spinning disk windows and engineer's control stand got a "Cadillac" nickname from maintenance crews, because of their special equipment and use only for snowfighting. As they get hooked up in the blizzard, brakeman Dale Baltazar gets ready to move them out. (Below) Because of the immobilizing effect of the deep soft snow, nose-shrouded "War Machine" No. 4839 and its flanger stops at the Gap to provide "commute service" for a Ray Chavez gang going down to dig out switches at the balloon track.

With "the pedal to the metal", a westbound hotshot thunders past the double spreader set in the blizzard above Truckee, keeping up its valuable momentum despite a growing snowpack on the 1.8% grade.

(Right) On a return pass, Jim Mahon, assistant division engineer, sends a wide swath of snow flying over the embankment at Upper Cascade. "The Bear" is a legend on this pass for his vigorous personal management of snow fighting activities.

SP

(Left) Widening out a fresh 6-foot snowfall on the No. 2 track at Norden, Charlie Drinnon's spreader set is overtaken on the No. 1 track by this morning's westbound *Zephyr*, giving Amtrak passengers the privilege of seeing Donner snowfighting equipment in full action. (Above) Drinnon's crew rushes to changes angles of the spreader's giant noseplates during a rare bit of sunlight at the Gap. (Left) At Norden, camp cook Marty Harris can still stand on the 15-foot snow measuring pole.

(Far Left) Even 500 tons of spreaders and locomotives dwindle to insignificance in the fury of a Sierra Nevada storm. We see the effect at tunnel 3 above Cisco in 1983 where a returning spreader set is almost swallowed up by the immense snowfall. (Below left) The summer view of the scene presents an incredible contrast. (Above left) As the afternoon wears on, Drinnon begins to frost over as he intently studies the buried right of way for clues to the hidden obstacles that could wreck the spreader's wing.

Night at Truckee finds a hot westbound freight hitting the mountain grade past the assembled spreader set, with its Mars light searching the dark sky for more falling snow. "Old reliable" Cadillac No. 4319 pauses outside the depot after one of its flanger "flips" to the Gap. Inside, Jim Mahon gets his train crew together to plan their next moves.

(Left) Its March,1983, and the spreader crew pushes on into the night through an incredible ocean of snow, swirling around the noseplates and over the huge wings of the spreader as a great flood to test the strengths of men and their machines.

DICK DORN

In spring of 1982 a total of 194 inches of snow fell in seven days at Norden on top of a 97-inch pack. This closed the railroad for five days, creating a week of extraordinary snow conditions on the mountain, with several rotary sets engaged fully in the battle to reopen the line. In the photographs by Dick Dorn, we see a moment's lull in the action at Norden as a monstrous wide-wing rotary stands idle while trainmen pause to talk over the situation during a crew change. (Right) Then the battle again is engaged in full, as SD45T-2 motors between the back-to-back rotaries blast slowly westward out of the Norden sheds with a smoke plume that would do justice to the steam rotaries and Cab-forwards of the previous generation. By Sunday, April 4, the Norden snowpack measured 19 feet and 9 inches, a bit lower than the record snowpack of 25 feet and 5 inches established in 1906-7.

(Right) In the latter stages of a major Sierra storm the rotaries act as clean up shovelers . . . grabbing everything in sight and tossing it out beyond the spreaders reach. This throwing of everything and anything is seen at Yuba Pass where a glint of sunlight suggests that the men and machines will soon be leaving the dark world of the mountain for one of light in the valley below. (Below) A variation of the same plan sees buried automobiles being yanked out of the drifts at Nyak before they perish like the frozen Mammoths in the Siberian Arctic.

DICK DORN

Amtrak
313
Amtrak

Eventually the storms cease and a final spreader run across the snow zone results in a clean railroad, as at Shed 10, where a long hotshot slowly passes a stopped spreader as Lupe Jimenez steps out to make an inspection. (Left) At the same spot outside Shed 10, Dorn's camera catches the morning's descent of the westbound *California Zephyr*. . . an exhilarating image of the Sierra in winter and the triumph of men over the elements.

The rotaries return to Roseville . . . until the mountain calls again.

Driving Donner

"Around Truckee it frosts any morning of the year, so bring your tire chains, boots and coat anytime you start driving up I-80."

In winter the best way to see Donner is to hop aboard the westbound Amtrak at Reno or Truckee and let the engineer take you up the snowy eastbound route to the summit. That dramatic climb, and the nearly 7000-foot descent to Roseville in the Central Valley, comprise one of the most interesting passenger train rides in North America.

In the longer days of summer, you can board the eastbound train at Sacramento or Roseville and still get a very real appreciation for the engineering and hard labor it took to push this railroad across the Sierra Nevada in the 1860's.

In case that doesn't fit your plans, let's take a trip across the mountain by car. You'll see the sights and find a few good photo locations without resorting to trespass or conflict with railroad safety.

By way of preparation, topographical sheets are okay but you'll need an AAA *Lake Tahoe Region* or *Feather River Region* highway map, or other map showing both Interstate 80 and the SP mainline.

Let's start where The Big Four started in 1863, and have breakfast at the Whistle Stop in Old Sacramento between Interstate 5 and the Sacramento River, before taking a tour of the California State Railroad Museum.

Have you had enough food and history? Let's head north on I-5 and go east on I-80 toward Roseville.

Unlike the perceptions of the state geologist in 1863, the Sierra Nevada range still seems to be safely in the distance as you pass Arcade Creek and cross the SP mainlines near Watt Avenue.

Take the Antelope Road turnoff in Roseville, going left until you see the SP tracks. Then turn right on Roseville Road and follow the edge of the yard. Continue on across the tracks on the Foothill Blvd. overpass and then turn off on the other side to get on the road paralleling the north side of the yard. Continue east past the auction yard to the far end of the SP yard where you see the diesel shops and service area. You can get a good view of the heavy motive power waiting for assignments, and perhaps a glimpse of the SP's fleet of flangers and electrified rotary snowplows. Rotaries are used only every few years or so, but you'll see them ready to roll in their double-ended configurations if things get tough on the mountain.

Use the underpass to cross below the yard, and follow the tracks east to I-80. We'll stay on the freeway to Newcastle today, but take Taylor Road if you want to check out the wandering main lines in the grassy hills above Rocklin and Loomis. At Rocklin the eastbound #2 track "flies" over the old westbound #1 track to start left hand running which lasts until Applegate. The eastbound grade to Colfax is a relatively gentle 1.5%, but the old westbound track has descents as steep as 2.4%. This was a big fruit shipping area for SP until orchards bloomed into subdivisions. But be careful in summer, the brown grass is a real fire hazard. Taylor Road and I-80 rejoin at Newcastle

At Newcastle on I-80, we pass beneath both mainlines before they split again to find their distinctively separate ways up the hill. As we enter Auburn we see the eastbound No. 2 track flying high above us on its way to Colfax.

At Bowman, a Road Conditions sign tells you whether or not you and your tire chains are going to get reacquainted. This is pretty much the cut-off point for fast food until you hit Truckee- so grab those cholesterol fries the Donner Party would have loved.

East of Bowman we cross both mainlines again as they wander through this Mother Lode country of Manzanita and Pines.

To see a little more of these hills, turn right at Applegate over the westbound main before turning left at the stop sign and passing under the eastbound grade. Note that right-handed operations have returned. There are a lot of interesting roads above Bowman, but as a veteran user of AAA towing services I can assure you that going off-road in wet weather in this red clay area is not a sane act.

As we meander along the eastbound rails we see one of the many tunnels of the newer #2 track, with its portal neatly outlined by hand-fitted local stone, one of very few tunnels with a small lake above the portal. The rails burrow beneath us before our road takes us back to the freeway at the Heather Glen interchange.

If you remember the 1910 Cab-forward photograph of Herb Sullivan, you'll love Colfax, which lies to our left as we turn off at that interchange on I-80.

As you drive into downtown you'll recognize Sullivan's scene— looking down on what once was the SP and the Nevada County Narrow-Gauge junction. Legend has it that the narrow gauge boys with their little teakettles loved to wait until a heavy SP passenger train started out next to them on the eastward grade. Then they'd yank out the old steamer's throttle and speedily overtake the SP train with its struggling Cab-forward. It helped that the NCNG line went downgrade sooner than the SP's. Today, helpers are still occasionally added to eastbound freights here.

Cross the tracks and go right. Pull up to a stop at Highway 174 and go left about a half-mile to the Cape Horn Bar where you can park on the right and look across the ravine at the eastbound track rounding distant Cape Horn. That rocky promontory was the first great challenge met by the indomitable Chinese construction crews, but historians still hotly argue whether or not they actually hung in baskets to blast the line into the face of that cliff above the American River. To the left of the Cape, the 1920's westbound line disappears into the "Panama Canal" tunnels.

As we head back toward Colfax we stop at the overpass across the SP, and see heavy power coming toward us from the west— a freight we must have missed while on I-80. Great, that gives us a train to follow.

It looks like the FROGM (Fresno-Ogden Manifest) what with a couple of PFE cars up front. The exhausts of her four big Tunnel Motors scour the overpass beneath our feet. Two more big units shove on the rear. A minute later we spot the lead units through the trees blasting up toward the Cape.

Let's head back through Colfax to get back on the freeway and drive below Long Ravine bridge, shown in many pictures with a little NCNG train passing beneath.

As we climb the grade toward Gold Run, we pass Magra and Secretown turn offs. Just about 1/2 mile beyond the latter, look ahead to your left and you may see two tracks curving to the left on a fill— the site of the old Secretown Trestle, made famous by the Alfred Hart picture showing it being filled in by Chinese laborers. In case you want to repeat the shot, forget it. After Hart got his classic shots, the pushcart brigade used his hilly camera platform for fill.

We could wait for the train here, but let's go on ahead and see some real pay dirt. Swinging off to the left at Gold Run, we have plenty of time, since the winding SP grade west of here is one of the tougher eastbound pulls. Go up across the tracks, and turn right to find a good photo location.

We're on what looks like a long high fill, with I-80 below us on the south and the Dutch Flat Diggings on the north. Actually this is a mile long remnant of an ancient gravel bed, blasted away on both sides by the last century's hydraulic mining. Local legend says this railroad embankment is so rich in gold that people have repeatedly offered to pay for a substitute steel trestle replacement for this last "piece of the rock".

Maybe, maybe not, but its a good place to toss a few rocks into the canyon before we stop and watch the scarlet and grey units roar past. In the days of iron men and woodburning locomotives, this was winter headquarters for the "Storm Kings", those very brave engineers who weren't afraid to notch their 4-4-0s all the way out to ram huge wedge plows through the drifts. Only after the first Leslie rotary plow arrived in 1888 could snowfighting begin to move away from such risky brute force tactics.

Let's continue east along the rails and go left on Ridge Road to Dutch Flat, where Judah wrote the Articles of Association for the Central Pacific on the counter of Dr. Daniel Strong's store. Doc Strong was the first person to identify the sets of ridges which could be used to carry a railroad to Donner summit, a find similar to that of Arthur Keddie who, years later, discovered a practical route through the forks of the Feather River for the Western Pacific Railway. In 1860 only two men knew a railroad could be built across the Sierra— Judah and Doc Strong.

With a tip of the hat to Doc, continue through Dutch Flat to join I-80 at the Alta turnoff. Go under the freeway and turn left on Casa Loma, the frontage road, which leads us up to the point where we turn 45 degrees down to the right as soon we see an SP overpass ahead. This narrow little road twists around to the right, over a bridge and the SP mainline, up a hill, through trees, to a spring and water trough on your left marking the old Casa Loma campground. Pull off to the right and park, even though you don't see any tracks. (If you start to cross the SP again, you've gone a little too far)

Now, walk over toward the canyon to see the SP mainlines above the several thousand foot deep North Fork valley. It's a great place for shooting eastbounds in the afternoon with the distant gorge in the background. We're near the old station of American where trains would stop to show passengers the view, but this place is called Rocky Point.

Our train got ahead of us, so let's go back on I-80 headed east. Turn left at the first interchange, Crystal Springs. Almost lost in the trees here is a highway classic— Old Baxter. Its your basic 1920's log cabin truck stop, hurt by only a few "improvements" over the years. Grab a couple iced teas and we'll jump back on I-80.

At the top of a long grade we pass the Blue Canyon turnoff, which you could take to see that old pre-1926 division point site from the days of single track. But let's hurry to try to catch our eastbound train at Yuba Pass.

Passing Nyak, Emigrant Gap and Yuba Gap interchanges, we pass over the SP and turn off on Highway 20 to Nevada City. Make the 180 degree turn and park on the other side of the road as soon as it has swung around to the west. This is the Yuba Pass area. We have time to get the cameras out for a shot as our train roars past, up toward the tunnels and rocks where the *City of San Francisco* was stalled in 1952. You'll need pictures to identify the exact spot, but this is where all heck was loose for three very snowy days that January. Hard to believe, but that little sandy hillside to the right of the tunnel had enough snow on it to knock over a standing rotary plow.

We roll back down onto I-80 and catch up with the train running up on the rocky mountainside above us. The railroad is pretty much inaccessible the rest of the way to Soda Springs so we just follow it as it runs along the granite ridges and through the tunnels way above us.

Tired of this? Okay, pull ahead and take the Soda Springs/Norden turnoff. Go right a mile on old US 40 to Soda Springs where a side road to the right crosses the tracks. Park here, it's a good safe place to photograph the railroad in action in all seasons.

The engineer gives us a wave as the four big units and rear end helper roar past us on the approach to the sheds at Norden. Since he's taking the direct two-mile tunnel route under Mt. Judah, there's no way we can catch him again before Truckee. So let's take it easy and continue up old US 40 past the site of old Norden. Looking down the steep bank to the right you can see the remaining snowsheds. Past Norden House, we see the No. 2 eastbound track starting its descent to tunnel 41, and the No. 1 track still climbing toward tunnel 6, located in the snowshed across the highway from Donner Ski Ranch. It was a quarter-mile of granite, and a bitch to dig with hand tools.

We drive on past the shed entrance and on our right see the tunnel's capped center haul hole where two of the Big Four once stood and argued about a steam drill. The old highway ahead is open today so let's go over the summit and stop at one of the first turnouts.

This is an incredible view. On the granite hillside to our right we see the east end of tunnel 6 with its concrete snowshed, short tunnel 7, and tunnel 8 where the line burrows into Donner Peak. The No. 1 track we see is longer and steeper than the two-mile tunnel route and has tighter clearances so it sees a lot less traffic— mainly coal and grain trains. Below, along the highway, Indian petroglyphs can be found on some of the rocks. This was a traditional boundary between the the Washo Indians of Nevada and the Maidu of the west slope of the Sierra. A minute later we spot our eastbound train far away in the distance above Donner Lake.

Okay, let's head down the curvy old highway to Donner Lake and Truckee, approximating the route down which Crocker sledded three locomotives and 35 miles of track in the winter of 1867-68.

We see more of the No. 1 track snowsheds and tunnels before we reach Donner Lake and continue to the I-80 interchange on the outskirts of Truckee. The area south of here along Cold Stream Canyon is being protected from development. With a good, high car, sometimes its possible to enter to see SP trains looping around that great canyon.

We take I-80 to the main Truckee exit, and turn right into this interesting little mountain town. There are a number of good country restaurants from which we can choose— but pick a window seat so we can see any trains that thunder up through town. The spreaders and most of the flangers operate out of here during winter storms, which makes for very interesting scenes. The daily Amtrak trains also stop here.

Okay, let's get going again if we're going to see this railroad. Continue east on the main street and turn off on Glenshire Drive which follows the tracks. There are interesting views as the rails and river twist around and pass beneath our road. We lose the railroad for a while but regain it again in a very photogenic area near Boca where we can get back up on I-80 to continue east.

Now we follow the railroad on the freeway, spotting any number of interesting places to shoot trains. We cross over the tracks again at Verdi, an old Mormon settlement. The West's first train robbery happened here in 1870. The West's second train robbery, of the already looted train, happened the next day in Eastern Nevada to joyfully provide the media with an early "dumb crooks" story.

Just past Verdi, we take the West 4th Street exit into Reno, following the rails most of the way to Virginia Street. Here we go right, toward the tracks, and park our car to find out what's behind all those bright lights. Its always exciting to be in downtown Reno and hear a horn and gates going down just before a very dark and real SP freight train roars over the Virginia Street crossing in the midst of this neon-lit world of gambling.

Later, we can drive a couple of miles across town to the SP terminal at Sparks where photography is pretty much discouraged except when people are out on the platform for the daily Amtrak trains. Our solution: stay at the Sparks' Nugget hotel and ask for a room overlooking the yard. The food's good and it's just a short walk in the morning to see the arrival of the westbound *California Zephyr*.

Have a good trip, and stay safe.

SHIRLEY BURMAN

Amtrak No. 6 climbs the grade at Emigrant Gap past skiers returning on Interstate 80.